THE
Magnet
BOOK

Shar Levine and
Leslie Johnstone

Illustrated by Jason Coons
Photographs by Jeff Connery

Sterling Publishing Co., Inc.

New York

To Maurice Bridge, editor of *Business in Vancouver,* and to Andrew Phillips, Washington Bureau Chief of *Maclean's News Magazine,* for their encouragement, support and mentoring over the years, and for their uncanny ability to make a sentence into English.—S.L.

To the Kelowna Johnstones, Bud, Edie, Paige, and Pam, with much love and affection.—L.J.

To Emily: May the four winds blow you safely home.—J.C.

Acknowledgments

Thanks to Liz and Liz at Technokids, Vancouver, B.C., Canada
Our gratitude to Dr. David Hawes for his advice and assistance

Designed by Judy Morgan
Edited by Isabel Stein
Illustrated by Jason Coons. Photographs by Jeff Connery

Library of Congress Cataloging-in-Publication Data

Levine, Shar, 1953–
     The magnet book / by Shar Levine & Leslie Johnstone.
          p   cm.
     Includes index.
     Summary: Provides instructions for about thirty simple experiments exploring magnetism and electricity.
     ISBN 0-8069-9943-8
     1. Magnets—Juvenile literature. 2. Magnets—Experiments—Juvenile literature. 3. Magnetism—Juvenile literature. 4. Magnetism—Experiments—Juvenile literature. 5. Electricity—Juvenile literature. 6. Electricity—Experiments—Juvenile literature. [1. Magnets—Experiments. 2. Magnetism—Experiments. 3. Electricity—Experiments. 4. Experiments.] I. Johnstone, Leslie. II. Title.
QC757.5.L49—1997
538'.4'078—DC21   97-13987
                                                                                          CIP
                                                                                          AC

1 3 5 7 9 10 8 6 4 2

Published by Sterling Publishing Company, Inc.
387 Park Avenue South, New York, N.Y. 10016
© 1997 by Shar Levine and Leslie Johnstone
Distributed in Canada by Sterling Publishing
c/o Canadian Manda Group, One Atlantic Avenue, Suite 105
Toronto, Ontario, Canada M6K 3E7
Distributed in Great Britain and Europe by Cassell PLC
Wellington House, 125 Strand, London WC2R 0BB, England
Distributed in Australia by Capricorn Link (Australia) Pty Ltd.
P.O. Box 6651, Baulkham Hills, Business Centre, NSW 2153, Australia
Printed in Hong Kong

Sterling ISBN 0-8069-9943-8

Contents

Preface 4

Note to Parents and Teachers 5

Safety Note 6

List of Supplies and Equipment 8

PART ONE

Magnets & Magnetism 9

Get a Lode of This 10

Using a Compass 11

Magnetic Attractions 13

Round and Round 15

Playing Fields 16

Shields Up 18

Tan Lines 20

Field of Streams 22

Which Way Did It Go? 23

Chain Gang 26

Shake It Up 27

Promises and Inducements 30

Eye of the Needle 31

Due South 32

What a Wonderful World 34

I'm So Inclined 35

Hammerheads 37

Hot Stuff 39

On the Level 40

PART TWO

Electricity & Magnetism 42

The Tale of Hans Christian Oersted 43

Slim Pickin's 46

Going Round in Circles 48

Merry-Go-Rounds 49

Faster, Higher, Stronger 51

Two Are Better Than One . . . or Are They? 53

Hot Shoes 54

Aluminum Slidings 56

Electric Motor 57

Electric Generator 60

Pump Down the Volume 62

Sounds Like Teen Spirit 63

Magnetic Subs 64

PART THREE

Magnetic Magic & Games 66

Curling Iron 67

An Extremely Charming Snake 68

Gone Fishin' 70

Keeping Afloat 71

Clip Art 74

Receding Hairline 76

Glossary 77

Index 79

Preface

Have you ever thought how magnets affect your everyday life? When most people think of magnets, something which looks like a bar or horseshoe immediately comes to mind. But magnets are more than that. For example, the lines and numbers on the bottom of bank checks are made of magnetic inks, which can be read by computers. These inks are also added to most paper money to help distinguish real money from counterfeit bills. In some countries, magnets are used in coin sorting machines and in toll booths to identify money.

Magnets are found in computers, videocassette recorders, and compact disc players. But uses for magnets don't stop there: they are also used in medicine to help diagnose and locate brain tumors. Without magnets we wouldn't have motors or electric generators, either. *The Magnet Book* traces the history of magnetism from the ancient Chinese, Greeks, and Vikings to our own time. You can perform the same experiments people throughout the ages have enjoyed. You will also learn of the theories used to explain magnetism. You will discover how a piece of wire, a nail, and a battery can be transformed into an amazing device. Learn how a compass works, and how you can trick it into pointing in the wrong direction.

At the end of the book, there is a section that is just for fun: make your own magnetic games and toys, which will amuse and amaze your friends and family. No book on magnetism would be complete without experiments on electricity. Discover how magnetism and electricity are related. Don't worry about getting shocked. These activities are safe and exciting.

This book contains simple and safe experiments using everyday materials found around the house. After you do them, you may want to learn more about magnetism. Look in your local library or even on the Internet for additional interesting experiments or information. This book will change the way you look at magnets.

NOTE TO PARENTS AND TEACHERS

Few things are as magical to children as magnets. Magnets can be used by even the smallest of children. Young children can be amused for hours on end. They love to experiment, holding magnets near things in the house and seeing what will stick. With a few exceptions, noted in the safety notes, this is a fun and safe thing to do. Adults aren't immune to the attraction of magnets, either. It's amazing to watch grownups play with iron filing toys and make funny pictures with them after the children have gone to bed.

After you've tested your entire home for magnetic materials, what next? In this book, both children and adults alike can learn more about magnets. Activities are arranged so that the first time a new technique or idea is introduced, it is described in detail. In subsequent experiments, there is a reference to the experiment in which the concept was explained. Scientific words are printed in boldface type and defined the first time they are used; many also appear in the glossary at the

back of the book.

The Magnet Book is intended to teach children basic experiment and observation skills. The experiments can be expanded to become science fair projects or classroom activities. If you are a teacher, you may wish to set up stations for several experiments and have groups of students work their way through the activities. Encourage students to keep a journal of their observations. Ask them to research the different ways magnets are used in everyday life. See who can come up with the most unusual use for a magnet and who can find the most uses for a magnet.

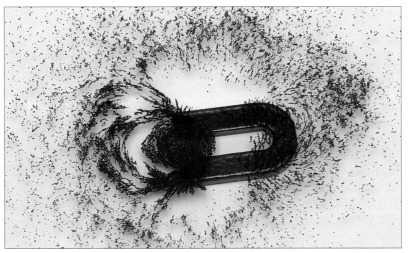

The most important thing is to have fun. These simple activities can be enjoyed by children of all ages. So break out the magnets and attract some children of your own!

SAFETY NOTE

Before you begin any of the activities in this book, there are a few things you should know about. Following these few safety tips will ensure that precious things such as your fingers, watch, and your computer remain unharmed after the experiments.

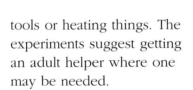

DO'S

1. Ask an adult before handling any materials or sharp tools or heating things. The experiments suggest getting an adult helper where one may be needed.

2. Connect batteries to electric circuits for short periods of time, and disconnect the wires if they begin to heat up.

3. Keep all supplies, tools, and experiments out of the reach of very young children, including small objects such as magnets, pins, paper clips, and batteries.

4. Dispose of used batteries in a safe way that doesn't harm the environment.

5. Read all the steps of any experiment carefully, assemble your equipment, and be sure you know what to do before you begin the experiment. Work on a stable surface where you have enough room.

6. Tell an adult immediately if you hurt yourself in any way.

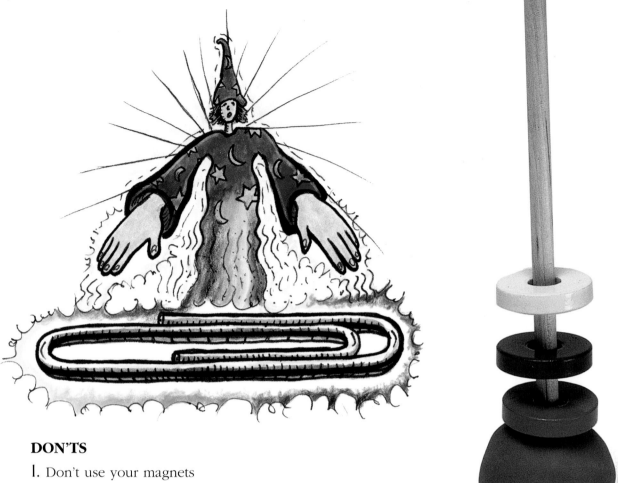

DON'TS

1. Don't use your magnets near computer equipment or software, or watches.

2. Don't use your magnets near television or stereo equipment. Keep magnets away from videotapes or audiotapes.

3. Don't open up batteries.

4. Don't put any tools or supplies in your mouth.

5. Don't leave your compass near a strong magnet for very long; it might demagnetize the compass needle.

SUPPLIES AND EQUIPMENT YOU WILL NEED

2 bar magnets, a horseshoe magnet, and a few circular magnets

compasses

sewing needle

2 or 3 corks from wine bottles

tissue paper

scissors

glue

masking and cellophane tape

steel and plastic-coated paper clips

iron filings or steel wool

insulated copper wire

6-volt lantern battery

string

modeling clay

aluminum foil

hammer

iron nails

craft knife

audiotape and audiocassette player

vegetable oil

wire coat hanger

iron bar

lodestone

tempera paints

plastic wrap

wooden sticks or bamboo skewers

thumbtacks or push pins

paper

pencils

felt pens or crayons

cardboard

wire stripper

acetate sheets

1.5-volt dry cell

tall, narrow glass or plastic jars

tweezers

steel knitting needles

wire cutter

straight pins

protractor

light-sensitive paper*

corn syrup

carpenter's level

sand

*Available from a hobby or toy store under the trademark Solargraphics®

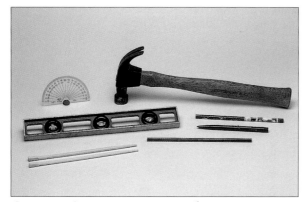

Some equipment: protractor, hammer, carpenter's level, straw, ballpoint pen, chopsticks.

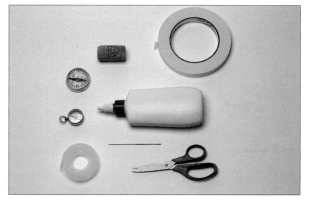

More equipment: compasses, cork, tape, white glue, needle, scissors, cellophane tape.

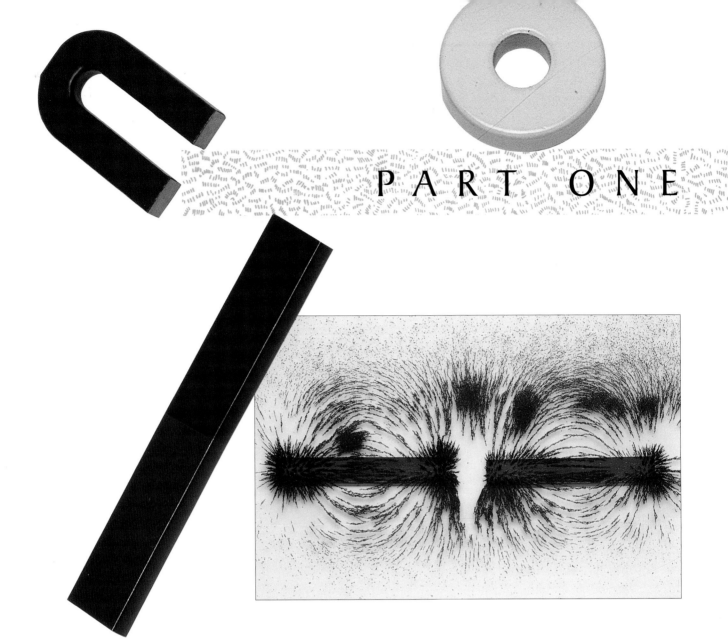

PART ONE

Magnets & Magnetism

Get a Lode of This

There is an ancient Chinese legend that tells a tale about Emperor Hwang-ti, who lived about 2600 B.C. It is said that during battle he was surrounded by a thick fog. He found his way through the fog by following a small pivoting figure with an outstretched arm, attached to his chariot. The figure always pointed south because it had a **lodestone** embedded in its outstretched arm.

Although we don't know if this legend is true, most scholars agree that the first compasses were made by the Chinese. The compass shown here is a reproduction of the earliest known compass; it was invented over 1600 years ago in China. The spoon was made of lodestone, a naturally magnetic rock composed of the mineral magnetite. When the spoon was placed on a bronze base, the handle of the spoon always pointed south. Here are just a few things you can do with a lodestone.

YOU WILL NEED

✔ lodestone or a manmade magnet
✔ several different kinds of sand
✔ different metallic and non-metallic objects, such as steel and copper paper clips, toothpicks, pins, needles
✔ paper and pencil

WHAT TO DO

1. Dip your lodestone or magnet into several different samples of sand. Do any sand grains stick to the lodestone?

2. Test the lodestone to see if it attracts other objects. Make a list of the objects that stick to the lodestone.

WHAT HAPPENED

A piece of lodestone will behave in the same way artificial permanent magnets do. The sand samples you col-

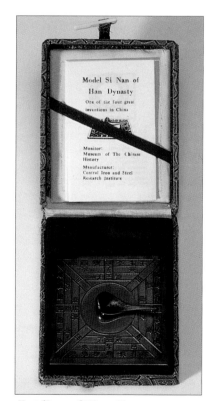

Replica of Han Dynasty Chinese compass.

lected may contain magnetite, which is found in many parts of the world, including Europe, Australia, and North and South America. Sand with black grains often contains some magnetite, although some black sands are made of mica, lava, or other rocks and minerals. If they contained magnetite, they were attracted to the lodestone. Some metal objects were attracted to the lodestone.

Using a Compass

If you were an adventurer exploring the jungles of the Amazon and you wanted to find out exactly where you were, you could simply turn on your laptop computer. You could download a satellite link; in seconds, a map showing you the area's latitude, longitude, and local terrain would be flashed before you. There would be no need for the old-fashioned **compass.** Until recently, explorers did not have laptops; they had to rely on compasses to guide them. Compasses are among the easiest scientific instruments to use; let's look at how they work.

YOU WILL NEED

✔ compass

WHAT TO DO

1. Place the compass on a flat surface, away from anything metal, and watch as the needle spins around. When

The needle of the compass is a small magnet, which is attracted by a very large magnet located right below your feet, the planet Earth. The Earth acts like a giant magnet, attracting magnetic material to its north and south magnetic poles. Scientists think the Earth's magnetic attraction may be caused by the movement of molten metals inside the Earth. Because the needle of the compass is a small magnet, it is attracted to some metallic objects. When you're trying to find a direction, use your compass away from metal objects and electric currents so you won't get a false reading. Some compasses have markings on them to show **degrees** (abbreviated °). A full circle is 360°. The degree markings can help you accurately find a location, if you know at what angle it lies in relation to north.

the needle stops moving, you will see that the colored end of the needle is pointing in one direction.

2. Turn the round compass casing so that the colored end of the needle is lined up with the letter *N* on the compass. The colored end of the needle points north, so you now know which way is north. If you face north, east *(E)* is to your right, west *(W)* is to your left, and south *(S)* is directly behind you.

WHAT HAPPENED

The compass needle turned to point towards the north.*

Actually, it points to the Earth's magnetic north pole, not the geographic north pole. The two are not the same. The difference in the angle between the two is called the magnetic declination, and often is printed on maps.

Magnetic Attractions

Why can a magnet pick up something as large and as heavy as a car, but not your little brother? When you hold a magnet close to an object, you feel the magnet pull towards the object, or the object moves towards the magnet. This is called **magnetic attraction.** The magnet attracts the object. If a magnet is pushed away from another magnet, the force is called **magnetic repulsion.** The magnet repels the object. Magnets are able to attract some materials; however, magnets have no effect on many other materials. Try this experiment to see which materials magnets attract.

YOU WILL NEED

✔ magnet
✔ steel paper clips, plastic-coated paper clips
✔ several small metallic objects such as coins, thumbtacks, push pins, and small metallic toys
✔ several small nonmetallic objects such as plastic toys, pieces of paper, corks, glass beads, and marbles
✔ several sheets of paper

WHAT TO DO

1. Place a pile of paper clips on a tabletop.

2. Hold the magnet over the paper clips. Notice how close you need to move the magnet to attract the paper clips.

3. Place the magnet close to the other metallic and non-metallic objects you have chosen. Which types of materials does the magnet attract?

4. To see which objects were attracted the most, place sheets of paper between each object that was attracted and the magnet. Try to pick up the object. See how many sheets of paper you have to place between the object and the magnet before the magnet stops picking up the object.

WHAT HAPPENED

The nonmetallic objects were not attracted to the magnet. (If a paper clip is steel or iron underneath a nonmetallic coating, you may have been able to pick it up, however.) Also, not all metals are attracted by magnets; for example, copper, silver, and gold are not attracted by magnets. Paper is not attracted by a magnet. The layers of paper that you placed between your objects and the magnet increased the distance between the two, which reduced the amount of attraction.

Metals that are strongly attracted by magnets include iron, nickel, cobalt, and alloys or mixtures of these metals, such as steel.

Some things were attracted by the magnet.

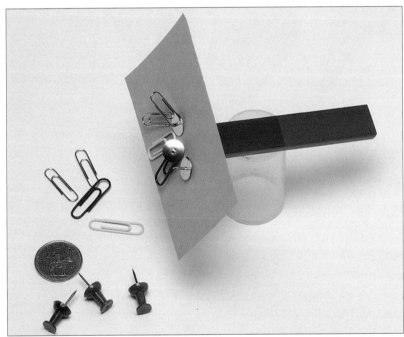

When a paper was placed between the magnet and the objects, some were still attracted; others weren't.

Round and Round

C an a magnet look for something? It turns out that the north end of a magnet is actually north-seeking and the south end is south-seeking (we call them "north" and "south" for short). The north-seeking end of the magnet is attracted to the Earth's north pole, because the Earth acts like a large magnet. The south-seeking end of the magnet is attracted to the Earth's south pole. Opposite ends of magnets are attracted to each other, while like ends of mag-

nets will repel each other. This experiment will help you better understand this principle.

YOU WILL NEED

- ✔ 2 bar magnets
- ✔ 8 inches (20 cm) stiff copper wire
- ✔ 1 yard (1 m) fishing line or string
- ✔ compass

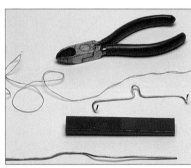

A wire cutter, string, a bracket made of wire, a bar magnet, a piece of wire.

WHAT TO DO

1. Make a holder for the bar magnet out of a stiff piece of copper wire. Bend the wire so that the ends support the magnet as shown in the photo.

2. Align your compass so you can tell which way north is. (See Using a Compass if you need help doing this.)

3. Attach the fishing line or string to the wire holder so that the holder is balanced in the middle and can move freely. Tape the free end of the string or line to the edge of a table or chair so that the holder can spin easily.

4. Center the magnet in the holder and allow the magnet to gently spin. Make sure that you are not standing near any metal objects. Wait for the magnet to stop spinning.

5. Look on your compass and compare the hanging magnet's direction with the compass's needle. The end of the magnet that points to the north is the north-seeking pole of the magnet (or simply, the north pole). The other end of the magnet is the south pole.

6. Take your second bar magnet. Bring one end near the north end of the hanging magnet. If the ends of the two magnets attract, the end of the second magnet nearest the hanging magnet is its north end. Turn the second bar magnet around. Is it now attracted or repelled by the hanging magnet?

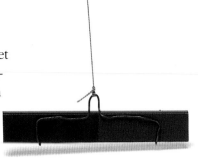

Opposite ends of the magnets are attracted to each other.

WHAT HAPPENED

The hanging bar magnet stopped spinning and pointed along the north–south line of the Earth's magnetic field. The end of the magnet that points to the north, the north-seeking pole of the magnet, is often labeled or stamped *N*. The end of the magnet that points to the south is the south-seeking pole of the magnet; it is often labeled or stamped *S*. In 1269 the French Engineer Pierre de Maricourt was the first person to write about the idea of magnetic poles. He discovered that when you place like magnetic poles together, they repel one another. The opposite poles, when placed together, will attract one another. Magnets cannot have fewer than two poles. At the poles, the magnet's magnetic force is strongest.

Playing Fields

In a book called The Invisible Man *by H.G.Wells, the police were able to find the title character by looking for his footprints in the snow. This was one way of making the invisible visible. Magnets are surrounded by invisible areas of magnetic effect called **magnetic fields**. This invisible effect can be revealed by sprinkling the area with small things that are attracted by the magnet. Here is a way that you can see the invisible.*

YOU WILL NEED

✔ two bar magnets
✔ horseshoe magnet
✔ clear acrylic or acetate
   sheet large enough to
   cover the magnets
✔ iron filings or steel wool
   cut into tiny pieces

WHAT TO DO

1. Lay the sheet of acrylic or acetate over the top of a bar magnet.

2. Sprinkle the sheet with iron filings or pieces of steel wool. (Be careful with the tiny pieces of metal; do not get them in your eyes.) Look at the patterns the filings make at the ends of the magnet.

3. Remove the magnet and return the iron filings to their container. Now try arranging the bar magnets end to end with the north ends facing each other but not touching (see photo). Place the plastic sheet over the magnets and sprinkle it with the filings or steel wool again. Look at the patterns the filings make between the magnets.

4. You can try a number of different arrangements with the magnets. Try them with a north end facing the south end, or place the magnets side by side. You can also look at the magnetic fields around a horseshoe magnet.

WHAT HAPPENED

The patterns that the filings created changed, depending on how you arranged the magnets. There were more iron filings at the two ends of the magnet, because the magnetic fields around your magnets are strongest at the ends. When you put the same poles near each other, north pole to north pole or south pole to south pole, there were fewer filings in

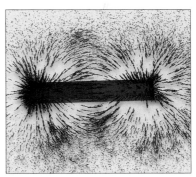

The magnetic field concentrated the iron filings at the ends of the bar magnet, where the field is strongest.

the space between the two magnets than when two opposite poles were near each other. When the opposite poles were placed near each other, they created a dense pattern in which the filings lined up between the

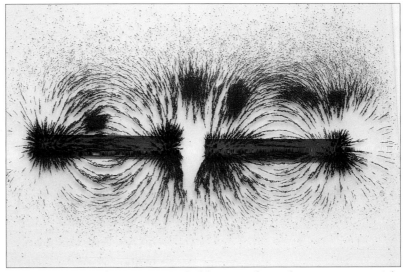

Iron filings in the magnetic field around two bar magnets with like poles facing.

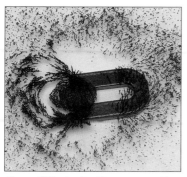

Iron filings in the magnetic field around a horseshoe magnet.

two poles. Like poles of magnets repel each other, or push each other apart, but opposite poles attract each other. The iron filings, which were magnetized, were pushed by the force of the magnetic field. The horseshoe magnet attracted most of the filings between its two ends. These ends are opposite poles, so the filings lined up between them.

If you were to break the magnets into shorter pieces, each piece would have a north pole that pointed in the same direction as the north pole of the original magnet. The south pole of the shorter pieces would point in the same direction as the south pole of the original magnet.

Shields Up

Thus far in this book you have been experimenting with magnetic attraction. What would happen if you wanted to keep a magnet from attracting an object? Is there a way for you to shield an object from magnetic attraction? In this activity, we'll see which materials reduce the attractive forces or block them altogether.

YOU WILL NEED

✔ magnet
✔ steel paper clips

✔ pieces of paper, cardboard, Styrofoam, aluminum foil, wax paper, leather, rubber, and cloth; a glass dish, pottery dish, and plastic plate or bowl
✔ iron or steel pot or pan

WHAT TO DO

1. Use one end of the magnet to pick up a paper clip. Try to pick up another paper clip with the free end of the first one. See how many paper clips you can connect in a row.

2. Remove the paper clips.

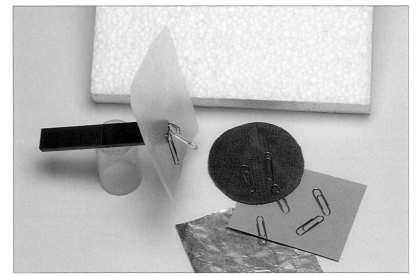

With a piece of paper between the magnet and the paper clips, fewer clips are attracted.

pottery or plastic over the end of the magnet before you try to pick up the paper clips.

4. Repeat step 2, but place an iron or steel pot or pan in between the magnet and the paper clips.

WHAT HAPPENED

The magnet was able to pick up paper clips through the piece of paper. You may have been able to pick up the paper clips with the magnet, if the other materials in between were not too thick. Magnetic fields can pass through nonmagnetic materials such as paper, glass, or plastic. Magnetic fields tend

With a steel pan between the magnet and the clips, no clips can be picked up by the magnet.

to change direction when they encounter magnetic materials such as iron or steel, so you couldn't pick up the paper clips if the magnet was placed on the far side of an iron or a steel pan from the clips.

Shake them up well to demagnetize them. Place a piece of paper over the end of the magnet and repeat step 1.

3. Repeat step 2, but instead of a paper, place a piece of cardboard, Styrofoam, aluminum foil, wax paper, leather, rubber, cloth, glass,

Tan Lines

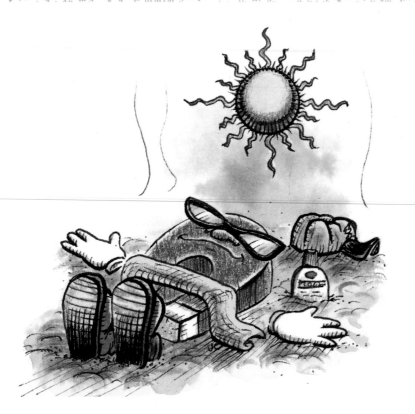

*A*t the end of the summer, do you have a tan line from your bathing suit? Do you think that magnets can leave tan lines, too? Here's a way to make a permanent record of the patterns created by iron filings using the sun and special light-sensitive paper such as Solargraphics,® available at hobby or crafts stores. You can create wonderful pictures to share with your class or to use in a science fair.

YOU WILL NEED

✔ magnets
✔ light-sensitive paper
✔ piece of cardboard slightly larger than the light-sensitive paper
✔ iron filings or small pieces of steel wool
✔ small pan or container larger than the paper
✔ water
✔ pen or pencil

WHAT TO DO

1. Place a magnet on the center of a piece of cardboard. Lay the light-sensitive paper, blue side up, over the magnet.

2. Sprinkle the iron filings on top of the paper. Carefully place the filings, magnet, cardboard, and paper in bright sunlight for about 5 minutes. If it is an overcast or cloudy day, leave them out in the sun longer. When the paper turns white, the picture is ready.

3. Take the filings, magnet, cardboard, and paper into a shady place. Lift the paper

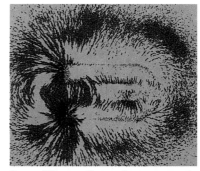

Iron filings show the magnetic field of a horseshoe magnet, which was underneath the paper earlier.

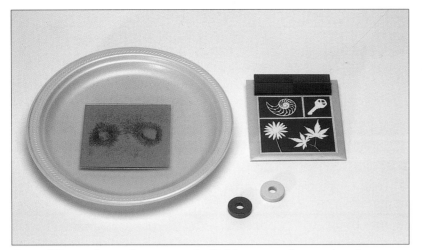

Iron filings show the magnetic fields around two circular magnets.

WHAT HAPPENED

The pattern of the iron filings showed the magnetic fields around the magnets. The light-sensitive paper changed color where it was exposed to the sunlight. The iron filings covered parts of the paper and didn't allow the sunlight to change the paper under the filings. The print made a record of the pattern that the filings had on the paper.

away from the magnet, and gently shake the filings back into their container.

4. Place the light-sensitive paper in a container of tap water and leave it there for several minutes. The paper should change from white back to blue.

5. Allow the paper to dry on a flat surface. Weight down the edges so the picture will not curl.

6. Try this experiment again, using different shapes of magnets or different arrangements of magnets on the paper. Make notes on the pictures, so that you remember how they were created.

After the filings are removed, the light-sensitive paper is developed, leaving a record of the magnetic fields around the circular magnets.

Field of Streams

W e've looked at magnetic fields on a flat surface and have seen the patterns the filings made. Is it possible to study the same patterns in three dimensions? In this experiment, different liquids are used to explore magnetic fields and the effect the type of liquid has on the field. Your creation can also make an interesting permanent sculpture after the experiment.

YOU WILL NEED

✔ 3 small jars with lids (baby food jars are perfect)
✔ iron filings or tiny pieces of steel wool (without detergent)
✔ vegetable oil
✔ white corn syrup or Karo syrup
✔ baby oil
✔ magnets: 2 bar, 2 circle, and one horseshoe

WHAT TO DO

1. Place a teaspoon of filings in the bottom of each jar.

2. Fill one jar with vegetable oil, one with baby oil, and one with white corn syrup. Close the jars tightly.

3. Shake the jars to mix the liquids with the filings.

4. Place a magnet with the north pole against the jar with the vegetable oil. Watch as the filings become attracted to the magnet.

5. Now place a magnet on either side of the jar, with the like poles (both north or both south) against the glass. How does the pattern change?

6. Switch one of the magnets, so the opposite poles of the two magnets are against the glass. Does the pattern change?

7. Use a horseshoe magnet to determine the pattern it will create with the filings.

A horseshoe magnet is placed to the side of one of the bottles.

8. Try these experiments again, this time using the other jars of liquid. Are the results the same each time?

9. To create a more permanent pattern, tape a small circular magnet to either side of the jar.

WHAT HAPPENED

The magnets attracted the iron filings in the jars. The filings were able to move more quickly through the thinner fluids such as the vegetable oil. When the opposite poles of the two magnets were placed near the glass, the filings lined up along the magnetic lines of force. When the like poles were across from each other, the repulsion of their magnetic fields caused the filings to move away from the area between the two magnets.

Which Way Did It Go?

When Robert Peary set out to find the North Pole in 1909 and Roald Amundsen searched for the South Pole in 1911, each needed special cold weather survival gear and scientific equipment. Just like the Earth, all magnets have a north pole and a south pole. What happens in between? What is the shape of the lines of force around a magnet? Here's a safe and warm way to go polar exploring with a compass and a magnet.

YOU WILL NEED

✔ bar magnet
✔ piece of paper
✔ compass or several compasses
✔ pen or pencil

WHAT TO DO

1. Place your compass alone on a table, away from your other equipment. Note which way north is (see Using a Compass if you need help).

2. Place your bar magnet down flat in the center of a

piece of paper. Put your compass between 1 and 2 inches (2.5 to 5 cm) from the north end of the magnet. Draw a small arrow on the paper, showing the direction that the north end of the compass needle points after it stops moving.

3. Move the compass around the magnet, each time recording the direction that the north end of the compass needle points. Take about 12 readings around the magnet. If you are curious to see how far the field of magnetic attraction reaches, move the compass further away from the magnet. When the compass needle isn't attracted to the magnet, you have found the outer edge of the magnet's range.

4. Place your magnet on its side, lengthwise, in the center of a new piece of paper. Move your compass around the magnet in the same way you did in steps 2 and 3; mark the direction in which the north end of the compass needle points.

5. Do the experiment again, this time with a fresh piece of paper and the magnet standing on its north pole. Then try it with the magnet standing on the south pole.

How did the compass readings change when you moved it around the magnet?

WHAT HAPPENED

The compass's needle aligned with the magnetic field of the magnet. If you connect the arrows, you will have drawn the magnet's magnetic **lines of force.** The

Arrows indicate the direction of the compass's north-seeking (colored) pointer as the compass was moved around the bar magnet.

The experiment is repeated with magnets standing on their poles. Left: with the magnet standing on the north pole, the north-seeking arrows are pointing away from the magnet. Right: with the magnet standing on the south pole, the north-seeking arrows are pointing towards the magnet.

compasses placed around the magnets showed that the magnetic lines of force move outward from the north end of the magnet and continue along the sides of the magnet. About halfway down the side of the magnet, the lines of force point outward, and then they move inward and go in towards the south pole of the magnet. This occurred when the magnet was placed flat on the table or on its side. When you placed the compasses around the magnet that was standing on its north pole, the north (colored) arrow of the compass needle pointed outwards, away from the magnet. When the magnet stood on its south pole, the north end of the compass needle pointed inwards, towards the magnet.

PROPERTIES OF MAGNETIC LINES OF FORCE

1. Lines of force never cross

2. Lines of force form complete circles or loops

3. Lines of force repel or move away from each other

4. Lines of force tend to contract, like elastic bands

5. Lines of force move from the north pole to the south pole of the magnet

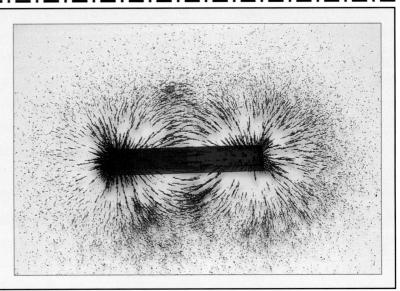

Chain Gang

B ar magnets have a rectangular shape. Horseshoe magnets are shaped like a U and resemble a horseshoe. Have you ever wondered why they are shaped in a curve? The answer lies in the next experiment. You will also discover if it is true that "two's company, three's a crowd." Here's a way to see not only which part of the magnet is the strongest, but also how many items a magnet can attract in one spot.

YOU WILL NEED

✔ bar magnet
✔ horseshoe magnet
✔ steel paper clips
✔ copper wire holder for your bar magnet (see Round and Round)
✔ nylon fishline or strong thread

WHAT TO DO

1. Center your bar magnet in a wire holder as shown in the photo. Hang your magnet's holder on fishline or strong thread. Allow one end of the magnet to pick up one paper clip.

2. Hold another paper clip just below the first clip and see if it is attracted.

3. Keep adding paper clips in this way until the last clip falls off. Depending on the strength of the magnet and the weight of the clips, you may be able to hang several clips from the magnet.

4. Try attaching a paper clip next to the first paper clip,

but closer to the middle of the magnet. Repeat steps 2 and 3.

5. Keep attaching more paper clips further and further in on the magnet, until you reach the center of the magnet. See how many paper clips the magnet will attract at its center.

6. Try this experiment again using a horseshoe magnet and see if the results are the same. (You can just tie a thread or fishline to the center of the U to hang the horseshoe magnet.)

WHAT HAPPENED

The magnetic attraction was strongest at the poles of the bar magnet. Each time you moved in from the poles of the magnet, fewer paper clips were attracted. Somewhere around the middle, the magnet attracted the fewest paper clips. The ends of the horse-shoe magnet attracted more paper clips than the middle of the horseshoe magnet. When both ends of a magnet are near each other, the magnet is more powerful than if they are far apart. Horseshoe magnets can pick up more in their curved shape than if they were straightened into a bar shape.

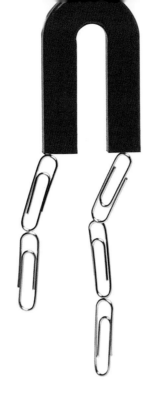

Shake It Up

To most people, iron filings simply look like tiny bits of black metal. It's a surprise to learn that each one of these filings can become a miniature magnet. You would think it would take forever to magnetize each and every one of these filings! Not only is there a simple way to magnetize filings, but you can also learn about magnetic domains at the same time.

Magnet, compass, filings, and tube or straw.

YOU WILL NEED

✔ thin plastic tube such as a toothbrush holder or candy container, or a wide clear drinking straw
✔ iron filings
✔ modeling clay to seal the ends of the plastic tube (if they are open)
✔ bar magnet
✔ compass

WHAT TO DO

1. Plug up one end of the thin plastic tube or straw with modeling clay, if it is not already closed. Pour in enough iron filings to fill the tube about three-fourths full.

2. Use modeling clay to seal the other end of the plastic tube, or use the tube's lid, if it has one.

3. Stroke the tube from one end to the other, using one pole of a strong magnet. Always stroke in the same direction (for example, from left to right). Be careful not to shake the tube.

4. Bring the tube near a compass. If the filings have been magnetized, the tube should attract one end of the compass needle and repel the other end.

5. Shake the tube so that the iron filings have a chance to be rearranged. Bring the tube near the compass again. What happens to the compass needle now?

WHAT HAPPENED

Magnetic materials can be thought of as being made up of a number of tiny magnets, each having its own north and south poles. Groups of these tiny magnets line up together to form small regions called **magnetic domains.** Within each

Magnetic domains. Top: in a magnetized piece of metal. Bottom: in an unmagnetized piece of metal.

region, the atoms are polarized (oriented) in the same direction. To make a magnet, we have to line up all or most of the domains in the same direction. When you stroked the tube of filings with the magnet, you lined up the magnetic domains of the iron filings. The filings became a temporary magnet and attracted the compass needle. When you shook up the filings, the filings became randomly arranged again, and attracted both poles of the compass needle.

WHAT CAUSES MAGNETISM?

When we take a piece of iron and make it into a magnet, we have changed something about the iron. What changed? Scientists tried to find the answer to this question for a long time. They found the answer by looking at the basic particles that make up matter.

All materials are made up of tiny particles called atoms. Atoms of iron are different from atoms of nickel, but are more or less the same as all other iron atoms. If we could look at the atoms that make up iron, we would see that they are made up of even smaller particles. Atoms have a nucleus or core made up of particles called protons and neutrons.

Charged particles called electrons move around the nucleus, like bees around a beehive. When electrons move, they produce small magnetic fields. Usually, electrons come in pairs. These pairs move in opposite directions and their magnetic fields cancel each other out. A piece of iron has a large number of unpaired electrons. In a piece of iron, the atoms align their magnetic fields with those of nearby atoms to form small magnetic areas called domains. When the domains are arranged randomly, a piece of iron or other magnetic material is not magnetized. When the magnetic material is placed in a strong magnetic field, the domains line up. The north poles of the domains all face in one direction, and the material acts as a magnet.

Promises and Inducements

I s it possible to make a metal object into a magnet without having another magnet touch it? Yes! This simple experiment will look like magic, when you get the hang of it.

YOU WILL NEED

✔ steel paper clips
✔ bar magnet
✔ soft iron nail
✔ paper

WHAT TO DO

1. Place some unmagnetized paper clips on a piece of paper.

2. Dip the pointed end of the nail into the paper clips. No paper clips should stick to the nail.

3. Dip the pointed end of the nail back into the paper clips, but this time hold the south pole of the bar magnet slightly above the flat end (head) of the nail.

4. Keeping the magnet the same distance from the flat end of the nail, lift the nail from the paper clips. Some paper clips should now be stuck to the nail.

5. Take the magnet away from the head of the nail; some of the paper clips should still be attached to the nail.

6. Without moving the nail, turn the magnet over so that now the north pole is held slightly over the head of the nail. What happens to the paper clips?

WHAT HAPPENED

The nail was not originally magnetized. This is why it did not attract any paper clips when it was originally placed near them (step 2). When you held a magnet over the nail, you caused the domains of the metal in the nail to line up. The nail became a temporary magnet. This is called **induced magnetism.** When you removed the bar magnet, the nail didn't lose all its magnetic abilities. It kept some of its magnetism; that is why the paper clips didn't fall off when the bar magnet was removed. This retained magnetism is called **residual magnetism.** When you turned the bar magnet so the north pole was nearest the nail, all the paper clips fell off the nail, because you reversed the poles of the nail. The paper clips also had some residual magnetism. They originally were attracted to the nail at their ends that had an opposite charge to that of the near end of the nail. When the poles of the bar magnet were reversed, the paper clips and the closest end of the nail had like poles, so they repelled each other and the paper clips fell off.

Eye of the Needle

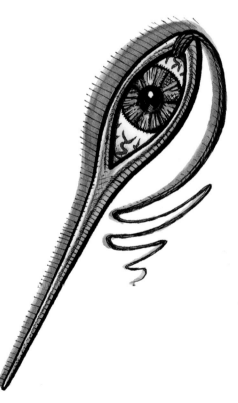

There are three kinds of magnets: naturally occurring magnets, such as lodestone (magnetite); artificial magnets, which are manufactured, and may be permanent or temporary; and electromagnets, which are temporary magnets made using electricity. Here is a fast and easy way to make a small magnet. When you finish, save your magnetized needle. You will need it in a few other projects.

YOU WILL NEED

✔ sewing needle
✔ steel paper clip (or plastic-covered steel paper clip)
✔ bar magnet

WHAT TO DO

1. Try to pick up the paper clip by touching it with the sewing needle. What happens?

2. Stroke the magnet along the length of the needle several times from the eye of the needle (the part where the thread goes) to the point. Do not rub the magnet back and forth across the surface of the needle.

3. Try again to pick up the paper clip with the sewing needle. What happens now?

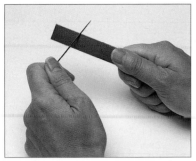

Stroke the magnet along the length of the needle from the eye to the point.

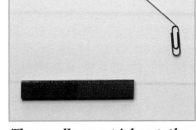

The needle now picks up the paper clip.

WHAT HAPPENED

The first time you tried to pick up the paper clip with the needle, it didn't move.

When you rubbed the needle with the magnet, you created a temporary magnet. The magnetic domains of the steel in the needle all became aligned in the same direction, and so the needle was magnetized. When you tried to pick up the paper clip with the magnetized needle, it was attracted.

Note: A needle can be magnetized by moving the magnet across it in a back-and-forth motion. We suggest stroking the needle with the magnet in one direction only, because it makes the needle magnetic faster, and it makes a stronger magnet.

Due South

The Vikings used a lodestone to help them find their way on the ocean. They floated the lodestone in some water to make a simple compass. With the lodestone, the Vikings were able to navigate their way to and from far-off lands, including North America. The word magnetite *comes from the Greek word* magnes, *meaning "a stone of*

Magnesia." Large deposits of these magnetic stones were found near the city of Magnesia in Asia Minor in ancient times. Magnetite, which is about 75% iron, is thought to have become magnetized by the magnetic field of the Earth. Magnetite was formed deep in the Earth. When it cooled, the particles of the mineral became locked in the same direction as the magnetic field of the Earth. Here's a way to make your own compass.

YOU WILL NEED

- ✔ nonmetallic bowl
- ✔ water
- ✔ magnetized sewing needle (made in the Eye of the Needle project)
- ✔ tweezers
- ✔ small slice of cork, like a wine cork, about one-eighth inch (3 mm) thick
- ✔ compass
- ✔ dishwashing liquid (optional)

WHAT TO DO

1. Place the bowl of water on a flat surface away from any metal objects.

2. Carefully lower the magnetized needle onto the surface of the water, using the tweezers. If you are unable to get the needle to float on the surface, try pushing the needle gently through a slice of cork (see photo); then lower the cork onto the surface of the water.

3. Wait for the needle to stop moving. If it is magnetized, the needle will point in a north–south direction. You have made a compass. You can check the accuracy of your homemade compass with a store-bought compass.

4. If the cork and the needle tend to float towards the edge of the bowl, add a few drops of detergent to the water, which will prevent this from happening.

WHAT HAPPENED

When you floated the magnetized needle on the water, the water allowed the needle to move freely. The needle

A floating cork slice pierced by a magnetized needle is a homemade compass.

became aligned with the Earth's magnetic field. The water just allowed the needle to move freely.

Some store-bought compasses use a water-filled chamber to allow the compass needle to move. Other compasses have needles that turn on a central pin. All compasses have magnetic needles, so they are affected by metals that might be nearby, sometimes even the user's watch or jewelry.

What a Wonderful World

In 1600, William Gilbert, an English doctor, proposed that the Earth was like a giant magnet. He had been experimenting with round pieces of magnetite and magnetized needles when he realized that the magnetite was attracting the needles in a way similar to the Earth's attraction of a compass needle. This is a way to make a model of the Earth and to explore its magnetic attraction.

YOU WILL NEED

✔ small bar magnet
✔ modeling clay
✔ paper clip
✔ string
✔ masking tape
✔ glass jar with a lid; it must be big enough to contain bar magnet
✔ light oil to fill jar, such as cooking oil, mineral oil, or baby oil
✔ iron filings

WHAT TO DO

1. Wrap modeling clay around a small bar magnet to completely cover the magnet and form a ball. Mark where the south pole of your magnet is on the clay before you cover it up.

2. Tie a piece of string onto a paper clip. Stick the paper clip into the ball of clay so that it is above the south pole of the magnet.

3. Tape the free end of the string onto the inside of the jar lid; leave a long enough piece of string between the lid and the ball of clay so the ball of clay will hang in the middle of the jar when the lid is screwed on.

4. Fill the glass jar with light oil, but not up to the top; leave enough room for the liquid to rise when you lower the clay ball into the jar.

5. Sprinkle about 2 tablespoons (30 mL) of iron filings into the oil, gently lower the clay ball into the jar, and seal the lid. Do not shake or disturb the jar. Allow the filings to gradually move into position around the clay ball.

WHAT HAPPENED

The iron filings were attracted to the magnet inside the clay ball. They were able to move through the oil and hang suspended near the poles of the magnet. The magnetic field of the Earth works in the same way; there is greater magnetic attraction near the poles. Each magnetic pole of the Earth is about 1000 miles (1600 km) from the geographic pole; the geographic poles are the points around which the Earth rotates. (The Earth's magnetic north pole is actually like the south pole of a magnet; it attracts the north poles of other magnets and repels the south poles of other magnets.)

The magnetic field of Earth.

I'm So Inclined

We generally associate the magnetic field of the Earth with a north–south direction. But did you know that the Earth's magnetic attraction also causes magnets to point down at an angle? The angle with the horizontal is called the **inclination.** At the magnetic north pole, the north-seeking needle of a compass would point straight down. In this experiment, we will construct a scientific instrument called a **dip needle,** which we can use to show that magnets do not only point north, but down as well.

YOU WILL NEED

- ✔ long, thin steel knitting needle
- ✔ cork from a wine bottle
- ✔ large metal sewing needle
- ✔ two identical drinking glasses or glass jars
- ✔ magnet
- ✔ plastic protractor
- ✔ cellophane tape

Equipment for the experiment: glasses, protractor, cork, knitting needle, sewing needle, and bar magnet.

The angle the dip needle makes with the horizontal can be measured on a protractor.

WHAT TO DO

1. Push a knitting needle through the length of a cork at its center. You might need an adult's help to do this. If necessary, drill a hole for the knitting needle first.

2. Have an adult push a sewing needle crosswise at right angles to the knitting needle, through the middle of the cork. It should stick out of the cork on either side (see photo). Make sure the needles you are using are not

A knitting needle and sewing needle stuck through a cork.

magnetized. You have just made your dip needle.

3. Place the dip needle between two glasses, so the sewing needle is on the rim of the glasses and the knitting needle can swing freely between the glasses.

4. Carefully adjust the knitting needle so that it balances evenly between the glasses and lies parallel to the top of the table.

5. Carefully remove the dip needle from between the glasses. Stroke the knitting needle with a magnet from the cork outwards. With the opposite end of the magnet, stroke the other end of the knitting needle from the cork outwards. Do not move the

knitting needle in the cork. Your knitting needle now has a south pole and a north pole.

6. Put the dip needle back between the two glasses as it was in step 3. Does the needle still stay parallel to the table?

7. If you wish to measure the exact angle of inclination, tape a plastic protractor to the far side of one of the glasses, away from the needle (see photo). Then you can read the angle between the horizontal and where the needle points to when attracted by the Earth's magnetic field, the inclination.

WHAT HAPPENED

After the dip needle was magnetized, it began to point to the table at an angle. The angle to the horizontal of the Earth's magnetic field varies, depending on where you live. At the "north" magnetic pole (in Canada), the dip needle points almost straight down (90° from the horizontal). Near the equator, the dip needle will stay parallel to the table (0° from horizontal). If you live between the equator and the pole, the angle of inclination will be something between 0 and 90°. At the "south" pole in Antarctica, the dip needle points up.

Hammerheads

Y*ou have discovered how easy it is to create a magnet from a piece of steel, using another magnet (see Eye of the Needle). The reason the steel became magnetized was that the poles of the magnetic* *domains of the steel were attracted to the opposite pole of the magnet, so the domains all ended up facing in the same direction. Is there a way to use an even bigger magnet to do the same thing? Here is an experiment in which you use the Earth to cause a piece of metal to become a magnet.*

YOU WILL NEED

✔ iron rod or large spike
✔ compass
✔ hammer
✔ steel paper clip
✔ protractor

WHAT TO DO

1. Place your compass on a flat surface away from other magnets or pieces of metal. Look at the compass needle. Align your iron rod or spike with the magnetic north and south, as the compass needle indicates. Hold the spike firmly so that it points in the same direction as the needle, and angle the rod downward at the local angle of inclination (see I'm So Inclined to find this angle). You can use a protractor to measure this angle.

2. Tap gently on the end of the rod or spike with a hammer several times. You may need someone to help you

with this if you find it difficult to hold the spike at an angle and hammer at the same time. Watch that you don't hit your fingers with the hammer.

3. Test your new magnet by bringing it near both ends of the compass needle. Then test your new magnet by using it to pick up a paper clip.

directions. When you struck the bar, you disturbed the domains of the metal inside the bar; then they became aligned with the magnetic field of the Earth. The magnetic domains all lined up in the same direction. The bar became weakly magnetized;

it will attract one end of the compass needle while repelling the other. It will also attract the paper clip. This is relatively easy to do with a soft iron rod, but becomes more difficult if you try to use a rod made of steel.

WHAT HAPPENED

The iron rod became a magnet. In an unmagnetized iron rod, magnetic domains face randomly in many possible

The magnetized rod picks up a paper clip.

CARING FOR YOUR MAGNETS

Magnets can lose their magnetism if they are exposed to strong magnetic fields or if they are struck or dropped repeatedly. They can also become demagnetized if they are heated to high temperatures. When this happens, some of the domains in the magnets lose their alignment and face in different directions and the magnets become less powerful; they may even return to being unmagnetized pieces of metal. Always store your magnets correctly, and be careful not to drop or bang them. Bar magnets should be stored in pairs, with the north pole of one next to the south pole of the other. Horseshoe magnets should be stored with a piece of unmagnetized iron placed across the end.

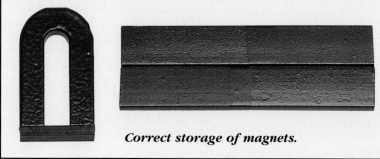

Correct storage of magnets.

Hot Stuff

*T*he ancient Chinese made magnets by heating iron bars to very high temperatures, aligning them in the Earth's north–south direction, and letting them cool. Many materials can only form magnets at very low temperatures. A few, such as iron, steel, cobalt, and nickel, can become magnets at room temperature. Let's see what happens to a magnet when it is heated.

YOU WILL NEED

✔ magnetized needle from the Eye of the Needle project
✔ candle
✔ matches
✔ pliers
✔ compass

WHAT TO DO

1. Using the pliers to hold the needle, place one end of the needle near the compass.

Does the needle move the compass point? If it is magnetized it should attract one end of the compass needle and repel the other end.

2. Have an adult perform this step. Light the candle. Using the pliers to hold the needle, place the needle in the flame of a candle. Hold the needle there for a few seconds. Make sure not to touch the needle, as it will be very hot.

3. Do step 1 again.

WHAT HAPPENED

When the needle was heated, the metal particles in the needle began to vibrate rapidly. The rapid vibrations allowed the magnetic domains to rearrange themselves randomly. When this happened, the needle lost its magnetism. (If the needle were lined up in the Earth's north–south direction, or if it were in contact with another magnet, it would cool to be magnetic, however.) Ferromagnetic substances such as iron, steel, cobalt, and nickel that have been magnetized will not remain magnetic if their temperature is raised above a certain point, called the Curie point. It is named after Pierre Curie, a French scientist who discovered it in 1894. The Curie point for iron is 770°C.

Holding a needle over the flame.

On the Level

By now you know that a magnet can attract iron objects and cause them to move towards the magnet. Did you know a magnet also can cause a bubble to move? In this experiment you will discover one of the stranger properties of magnets.

YOU WILL NEED

✔ carpenter's level
✔ strong magnet (bar or horseshoe)

WHAT TO DO

1. Place the carpenter's level on a tabletop. You will see a tiny bubble in the middle of the liquid in the level.

2. Move a strong magnet near the bubble and watch it move.

3. Use the other end of the magnet to see if the bubble still moves.

WHAT HAPPENED

The magnet made the bubble move. The liquid in the level is **diamagnetic.** This means that it is repelled by a magnet. Most substances are very weakly diamagnetic. When placed near a strong magnet, instead of being attracted they are repelled, and align themselves at right angles to a magnetic field. Bismuth, antimony, zinc, and glass are other examples of diamagnetic substances.

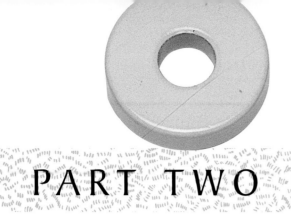

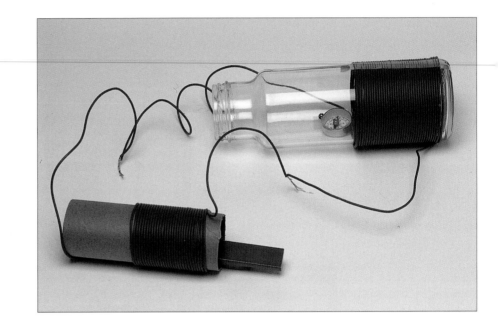

PART TWO

Electricity
& Magnetism

The Tale of Hans Christian Oersted

In 1820, a Danish scientist named Hans Christian Oersted made an amazing discovery. He found that when he placed a compass near a piece of wire that had electricity running through it, the compass needle changed direction. This showed that electricity and magnetism were related. Here is a way to repeat Oersted's discovery and to learn about an important rule.

YOU WILL NEED

- ✔ about 1 yard (1 m) of insulated (plastic-coated) copper wire
- ✔ large paper or Styrofoam cup
- ✔ wire stripper or craft knife
- ✔ 6-volt lantern battery
- ✔ small compass'
- ✔ felt-tipped marking pen

*It should be small enough to fit on the bottom of the upside-down cup.

WHAT TO DO

1. Place a compass on the table away from your other equipment. Note in what direction the north end of the needle points.

2. Remove about 1 inch (2.5 cm) of the plastic insulation from both ends of the piece of copper wire using a wire stripper, or have an adult remove the plastic insulation using a knife.

3. Poke a hole in the center of the bottom of the cup, and push the wire through the hole so that about half of the wire is sticking out.

4. Place the cup upside down on a tabletop, and attach the end of the wire that sticks out from under the cup's rim to the positive (+) terminal of the battery.

5. Put the compass next to the wire coming through the hole in the cup.

6. Attach the end of the wire that sticks up through the

Setup of equipment.

center hole in the cup's bottom to the negative (–) terminal of the battery. Look at the direction in which the compass needle's north (colored) end points. Use the marking pen to draw an arrow on the cup to show the direction of the north-seeking end of the compass needle at that place on the cup.

7. Move the compass around the wire, and mark the direction of the north end of the needle in each new position.

WHAT HAPPENED

When you connected the wire to the negative pole of the battery, you completed an electric circuit. An **electric circuit** is a continuous path through which electric charges flow. The flow is called an **electric current.** Just as Oersted did, you saw that an electric current creates a magnetic field. The current caused the compass needle to change direction from its original north–south position. As you moved the compass around the wire on the cup, you saw that the compass's north pointer changed directions. It followed the magnetic field of the wire, rather than the Earth's magnetic field. When you were finished you should have had arrows pointing in a counterclockwise direction around the wire.

Closeup showing orientation of compass's north-seeking needle when the current was flowing.

RIGHT-HAND RULE

You can use the thumb and fingers of your right hand to remember the direction in which the magnetic field travels around the wire. If the thumb of your right hand points along the wire in the direction the current is flowing (from positive to negative) towards the negative electrode of the battery, the magnetic field is moving around the wire in the same direction as your fingers curl outward around the wire. This is called the right-hand rule.

In this section of the book, many projects use a battery as a source of electricity. We use it to supply **direct current.** From the outside, batteries look mysterious, but the way they work is not at all mysterious. If you could look inside a battery, you would see that it is made up of two or more electrochemical cells joined together.

Electrochemical cells produce electricity by a chemical reaction. The cells have two main parts: a pair of electrodes or terminals and the **electrolyte.** The electrodes are made of two different materials, often carbon and a metal such as zinc. The electrolyte is a material that allows electrically charged particles to flow through it. The chemical reaction occurring in the cell causes the electrons to collect at the negative (–) terminal. When a conducting material like a wire connects the positive (+) and negative terminals of the battery, the electrons flow between them. This allows the chemical reaction to continue until the reacting chemicals are used up.

Don't leave the wires attached to both terminals for very long, or your battery will get used up. When this happens, the battery doesn't work anymore. When your battery is used up, you must take special care to dispose of it safely. Don't open batteries, as the chemicals inside the battery can harm you or the environment if they are released. A battery may explode if thrown in a fire. Only certain types of batteries can be recharged, so have an adult check your batteries before trying to recharge them.

Slim Pickin's

In the last experiment, we saw that electricity can create a magnetic field around a wire and move a compass needle. Let's see if the magnetic field of the wire is strong enough to pick something up.

YOU WILL NEED

✔ 6-volt lantern battery
✔ 2 to 3 feet (30 to 45 cm) insulated copper wire
✔ wire stripper or craft knife
✔ iron filings
✔ paper

WHAT TO DO

1. Have an adult trim the plastic ends from the insulated wire, so that several inches of copper are exposed.

2. Fold the wire in half. Where the wire bends, carefully expose about 1 inch (2.5 cm) of insulation by removing the plastic surrounding the wire. Make sure not to cut through the copper wire.

3. Place a small amount of iron filings on a piece of paper.

4. Attach one end of the wire to the positive (+) terminal of the battery; then dip the exposed center section of wire into the filings.

5. Attach the other end of the wire to the negative (−) terminal of the battery and dip the exposed center section of wire into the filings. Lift the wire above the paper. Did the filings stay attached to the wire?

6. Remove one end of the wire from the battery and observe what happens to the filings.

Dip the exposed center section of the wire into the filings.

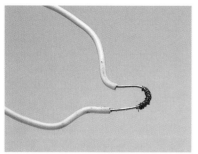

The wire attached to both terminals attracts the filings.

WHAT HAPPENED

The exposed wire only attracted the filings when it was attached to both terminals of the battery. The wire is made of copper, which does not normally attract iron filings. When the wire was connected to both terminals of the battery, an electric circuit was completed and current ran through the wire. The current created a magnetic field, which attracted the filings. When one end of the wire was disconnected from the battery, the circuit was broken and the filings fell off.

THE FLOW OF A CURRENT

At some time you may see a diagram of an electric circuit. In such diagrams, the rule (or custom) is that electricity flows from the positive to the negative terminal. This idea was thought up before people actually learned that electrons flow from the negative to the positive terminal. When we talk about the flow of current through a wire, we are talking about the direction from the positive to negative terminal, even though we now know that electrons actually move in the opposite direction.

Going Round in Circles

In Which Way Did It Go? we were able to map the magnetic field of a magnet by using compasses. In Slim Pickin's, we were able to pick up iron filings by generating a magnetic field around a wire. Now let's see what the magnetic field around the wire looks like, using those same filings.

YOU WILL NEED

✔ unpainted metal coat hanger
✔ wire cutter
✔ 4 × 4 inch (10 × 10 cm) piece of stiff cardboard
✔ modeling clay
✔ iron filings
✔ 6-volt lantern battery
✔ two 12-inch (30 cm) pieces of insulated wire

WHAT TO DO

1. With your wire cutter, cut a metal coat hanger apart just below the hook; cut off and discard the hook. Ask an adult to help you do this, if necessary. Bend the coat hanger so that it forms a stand (see photo).

2. Poke the hanger through the center of the cardboard, so that the wire extends several inches (5 to 10 cm) above the cardboard.

3. Place a small amount of modeling clay beneath the cardboard to keep it from sliding down the stand.

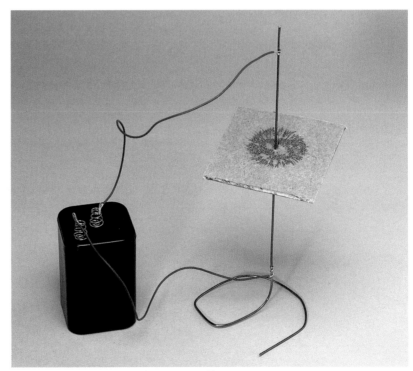

Equipment setup with current flowing.

4. Sprinkle iron filings on top of the cardboard.

5. Have an adult trim the plastic from both ends of each piece of the insulated wire, so that several inches of copper are exposed.

6. Wrap the end of one of the wires around the top of the stand and attach the other end of the wire to one of the battery terminals. Keep the extra wire away from the top of the cardboard.

7. Wrap one end of the other wire to the bottom of the stand and attach the other end of this wire to the free battery terminal.

WHAT HAPPENED

The filings formed circular patterns on the cardboard. These patterns are caused by the magnetic field that forms around the wire when there is an electric current moving through the wire. You attached the copper wire to the wire coat hanger, so the electrical current was able to move from the battery, through the copper wire, into the coat hanger wire, and back to the second piece of copper wire to complete the electric circuit at the battery. The magnetic field is stronger in some areas than in others; these areas, which show up as rings, attracted more of the iron filings.

Merry-Go-Rounds

YOU WILL NEED

✔ 6-volt lantern battery
✔ about 25 feet (8 meters) of insulated copper wire
✔ empty glass or plastic bottle or large round cardboard tube
✔ wire stripper or knife
✔ masking tape
✔ compass

At parties when people are blindfolded to play a game like Pin the Tail on the Donkey, you spin them around and around to confuse their sense of direction. This doesn't usually happen to a compass. It would be able to find north even if it were turned around. But sometimes it is possible to spin a compass and confuse it too! Here is an experiment that shows you how.

WHAT TO DO

1. Place a compass on the table away from anything magnetic or metal. Note what direction the north end of the needle points.

2. Remove the plastic from the insulated wire so that several inches of copper are exposed; get help from an adult if necessary.

3. Tightly wrap the wire around the bottle or tube so that it circles the tube at least 40 times. Leave about 1 foot (30 cm) of wire free at each end. You have made a **solenoid.**

4. Tape the coil to the bottle or tube to keep it in place.

5. Lay the compass inside the solenoid. Which way is the compass needle pointing?

6. Attach one end of the wire to the positive (+) terminal of the battery and the other end to the negative (−) terminal of the battery. Don't leave the wire connected for more than a few seconds at a time. Watch what happens to the compass needle.

7. Connect and disconnect the wires from the battery several times; each time, observe the compass needle.

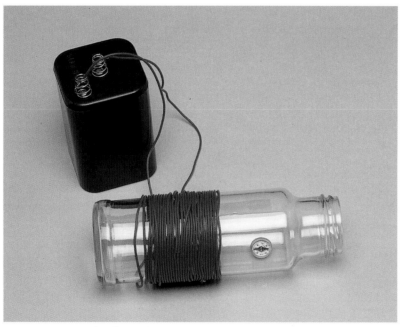

Compass inside the solenoid.

WHAT HAPPENED

When the wires were attached to the battery, the compass needle moved. Even though the copper wire alone is not a magnet, when you put an electric current through it, each coil of wire is turned into a miniature magnet.

The magnetic field outside the solenoid is like that of a bar magnet. The end connected to one terminal of the battery is the south pole and the end connected to the other terminal is the north pole. When you connect the battery, the compass needle changes direction from pointing to the Earth's magnetic north pole to pointing to the solenoid's temporary south pole. When you disconnect the wires, the compass needle once again points to the Earth's north pole. If a solenoid has an iron core inside it, it can form a very strong magnet, called an **electromagnet.**

Note: Keep your solenoid. You will need it for other experiments.

Faster, Higher, Stronger

Soon after Oersted's discovery of the relationship between electricity and magnetism, French scientist André Marie Ampère began performing a number of experiments based on Oersted's work. He found that when a wire carrying an electric current was bent into a loop, the magnetic force around it seemed to become stronger. Then he wound the wire into a coil and found that the coil acted like a bar magnet. He showed that a piece of iron inserted in the coil became strongly magnetized. Other scientists tried different ways to make magnets stronger; by the 1830s, they were able to make an electromagnet that could lift over a ton of iron.

YOU WILL NEED

- ✔ 10 feet (3 meters) of insulated wire
- ✔ wire stripper or craft knife
- ✔ 6-volt lantern battery
- ✔ large iron spike or steel nail or bolt
- ✔ steel paper clips
- ✔ 1.5-volt dry cell
- ✔ wooden pencil or chopstick

WHAT TO DO

1. Have an adult help you remove about 1 inch (2.5 cm) of insulation from each end of the piece of insulated wire. Attach one end of the wire to the positive (+) terminal of a 6-volt lantern battery.

2. Wrap the wire around a large iron nail so that the wire makes 10 turns around the nail. There should be a lot of wire left over on one end.

3. Attach the loose end of the wire to the negative (–) terminal of the battery. You have made an electromagnet. See how many paper clips the electromagnet will pick up. Do not leave the electromagnet attached to the battery for more than a few seconds; remove the wire from the negative terminal of the battery.

4. Wrap the wire around the nail so that the wire makes 20 turns and repeat step 3. Then try it again with 30 turns and with 40 turns, if you have enough wire.

5. Remove the wire from the positive terminal of the battery. Without unwinding the wire, reattach the wire to the terminals of a 1.5-volt dry cell. Now see how many paper clips the electromagnet will pick up.

6. Carefully slide the iron spike out of the middle of the coil of wire, and replace it with a wooden pencil or chopstick. Now see how many paper clips the electromagnet picks up.

WHAT HAPPENED

The strongest magnet you made was the one made using the 6-volt battery, the iron core, and the largest number of turns of the wire. The strength of an electromagnet is affected by the material used for the core, by the number of turns of wire, and by the amount of electricity flowing through the wire. Magnetic materials used for cores (for example, nails, bolts, or pieces of iron) make stronger electromagnets than nonmagnetic materials such as glass, wood, or plastic cores do. Increasing the number of turns of wire in the electromagnet also increases its strength. Each loop of wire has a magnetic field; when you place the loops next to each other, you create a more powerful magnetic field. Using more electric current will also increase the strength of an electromagnet. The greater the current is, the stronger the magnetic field becomes.

Here the iron spike was replaced with a chopstick.

Two Are Better Than One...Or Are They?

In Faster, Higher, Stronger, you made an electromagnet. Electromagnets can have different uses and different strengths. Some can pick up large objects, such as cars; others are used in small appliances, such as tape recorders. You know how to make an electromagnet stronger by increasing the electric current. Here's an experiment to show you why a current has to run in only one direction to power an electromagnet.

YOU WILL NEED

✔ 2 pieces of insulated wire about 3 yards (3 meters) long
✔ wire stripper or craft knife
✔ 6-volt lantern battery
✔ large iron nail or bolt
✔ steel paper clips

WHAT TO DO

1. Have an adult remove about 1 inch (2.5 cm) of the plastic insulation from each end of the pieces of insulated wire, using a wire stripper or a knife.

2. Attach one end of one of the wires to the negative (–) terminal of a 6-volt lantern battery. Wind the rest of the wire around a large iron nail or bolt from the top towards the bottom.

3. Attach the free end of the wire to the positive (+) terminal for a few seconds. You have made an electromagnet. Use your electromagnet to pick up as many paper clips as possible. Remove the wire from the positive terminal.

4. Attach one end of the second piece of wire to the negative terminal and wind the wire over the top of the first piece of wire. Make sure that you wind it in the same direction as the first piece of wire.

5. Attach the free ends of both wires to the positive terminal. Try to pick up paper clips. Is this electromagnet stronger or weaker than the one with one wire? Remove the wires from the positive terminal.

6. Now we're going to reverse the current in one wire. Remove the end of the second wire from the negative terminal and attach it to the positive terminal. Attach the free end of the first wire to the positive terminal. Attach the free end of the second wire (which previously was attached to the positive terminal) to the negative

Two wires are wrapped around the nail; both have the current running in the same direction.

terminal. Try to pick up paper clips. Is the magnet stronger or weaker than with one wire?

7. Remove the wires from the positive terminal.

WHAT HAPPENED

When you connected the first wire to the battery, you created an electromagnet, which could pick up paper clips. When you used two wires that had the electric current going in the same direction, in steps 4 and 5, the electromagnet was stronger than with one wire, and it picked up more paper clips. When the electric current went in one direction in one wire and in the opposite direction in the other wire (in step 6), the magnetic field in one wire was in the opposite direction to the field in the second wire, and they cancelled each other out. The wire-wrapped nail no longer acted as an electromagnet.

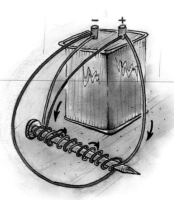

Top right: Both wires' electric currents go in the same direction. Bottom right: The currents go in opposite directions.

Hot Shoes

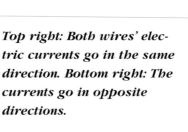

Electromagnets come in many different shapes and sizes; they are part of many devices that you see around you every day. Your doorbell, buzzer, telephone, tape deck, radio, and television speakers all contain electromagnets. Here is a way to make your own homemade horseshoe electromagnet.

YOU WILL NEED

✔ a large iron or steel U-bolt
✔ several yards (meters) of insulated copper wire
✔ wire stripper or craft knife
✔ 6-volt lantern battery
✔ steel paper clips

WHAT TO DO

1. Have an adult help you remove about 1 inch (2.5 cm) of insulation from one end of a piece of insulated wire. Attach the wire to the positive (+) terminal of a 6-volt lantern battery.

2. Hold the bolt with the curve at the top. Bring the wire behind the bolt. Starting at the lower left arm of the U-bolt, wind the free end of the wire around the left arm of the bolt towards the center, from the outside to the inside and back outside on the left. Continue to wind the wire around the left arm of the bolt in the same way as the first coil, moving upward towards the bend in the bolt. Put on as many coils of wire as possible. When you reach the curve in the bolt, bring the wire to the inside.

3. Then bring the wire around the back of the right arm of the bolt and start wrapping the right arm of the

bolt towards the left, from the outside to the inside, and then around the back to the right again. Continue to wind the wire on the right arm of the bolt in the same way as the first coil, moving downwards until you reach the bottom of the right arm of the bolt. The wire should be wrapped so that if you were to straighten the bolt, the wire would form one continuous coil, wrapped in the same direction.

4. Leave about 1 foot (30 cm) of wire sticking out at the end. Have an adult help you cut the wire and strip about 1 inch (2.5 cm) of insulation from the free end of the wire.

5. Attach the free end of the wire to the battery's negative (–) terminal; try to pick up the paper clips. Then detach the wire from the battery.

A horseshoe electromagnet.

WHAT HAPPENED

You made a horseshoe electromagnet. One of the ends of the electromagnet should be a north pole and the other end should be a south pole. If the ends have the same polarity, the wire was wound incorrectly. (If this happens, you just need to rewind one side of the magnet.)

Diagram showing how wire is wrapped around the U-bolt, starting from the positive (+) terminal.

Aluminum Slidings

The next time you flip a coin consider this: you are causing **eddy currents.** What is an eddy current? you may ask. As the coin rotates in the Earth's magnetic field, small electric currents form inside the coin in tiny loops. If a current flows in a loop, a small magnetic field is created. In this experiment, you will see how you can create eddy currents and use them to move a non-magnetic material with a magnet.

YOU WILL NEED

- ✔ aluminum foil
- ✔ scissors
- ✔ 2 pencils
- ✔ strong horseshoe magnet

WHAT TO DO

1. Cut a strip of aluminum foil so that it is 4 inches (10 cm) wide and 10 inches (25 cm) long.

2. Smooth out any wrinkles in the foil by placing it on a flat surface and pushing the wrinkles outward with your fingertips.

3. Place the two pencils side by side about 4 inches (10 cm) apart on a flat surface. Lay the foil over the top of the pencils so that about 3 inches (7½ cm) of foil sticks out on each side of the pencils.

4. Try to pick up the foil with the magnet. Does it move?

5. Rapidly move the poles (ends) of the horseshoe magnet back and forth just above the aluminum foil. The magnet should travel the full 10 inches (25 cm) of the foil, but should not touch the foil. Watch what happens.

WHAT HAPPENED

As you moved the strong magnet back and forth, it caused the aluminum foil to move. Aluminum is not usually attracted to a magnet as iron, nickel, or cobalt are.

The rapid movement of the magnetic field back and forth caused small eddy currents of electricity in the aluminum foil. These small electric currents produce magnetic fields. The pencils reduced the contact between the foil and the table, so that even a small force could move the foil. The magnetic fields that were induced in the aluminum foil were attracted to the magnet, and the foil was moved by this force of attraction.

Electric Motor

The next two experiments are based on discoveries made by the scientist Michael Faraday. Faraday, who was born in 1791, was the son of a poor blacksmith. He didn't attend school for many years. Instead he was apprenticed at a young age to a bookbinder in London, England. He was able to read many books and became very interested in science. In 1812 he attended some lectures given by Sir Humphrey Davy, a famous scientist. He sent Davy a copy of the notes he had taken at the lectures and asked Davy for a job. He then worked for several years as Davy's assistant. In 1821, he discovered the principles that allow electric **motors** to work. An electric motor is a machine for converting electrical energy into mechanical energy. Here is one way to make a simple electric motor.

YOU WILL NEED

✔ 8 long dressmaking pins
✔ large wine-bottle cork
✔ razor blade or craft knife
✔ about 4 yards (4 meters) of copper wire (insulated or uninsulated)
✔ wire stripper
✔ piece of rigid cardboard or balsa wood about 6 inches × 4 inches (15 cm × 10 cm)
✔ 2 metal paper clips
✔ 2 push pins or thumbtacks
✔ Two 1-foot (30 cm) pieces of insulated wire
✔ 6-volt lantern battery
✔ 1 large horseshoe magnet or 2 small bar magnets

WHAT TO DO

I. Have an adult use a knife or razor blade to cut two narrow grooves in the long sides of a large cork.

2. Push a dressmaking pin into each end of the cork to act as an axle. The cork should be able to rotate smoothly around this axle. The pins should stick out of the cork about ¾ inch (2 cm).

3. Have an adult remove about 1 inch (2.5 cm) of the plastic insulation from each end of the long piece of wire, if it is insulated. Wrap the long wire around the cork about 40 turns, fitting the wire tightly into the groove in the cork. Both ends of the wire should stick out about 2 inches (5 cm) from the one end of the cork.

4. Stick two straight pins into the end of the cork where the wire ends stick out. These terminal pins should be across the cut in the cork from each other and stick out of the cork about ½ inch (1 cm). Take one end of the wire and wrap it securely around one of these pins. Wrap the other end of the wire around the other pin. These will serve as the terminals for the electric current to enter the coil of wire.

5. Push two straight pins into a piece of rigid cardboard or balsa wood to form an X shape. Do the same thing a few inches away on the cardboard with two more straight pins. Place them so that your cork can fit between the Xes. Rest the axle pins sticking out from the cork in the V of the crossed pins so that the cork is suspended above the cardboard and can turn freely.

6. Bend one end of a paper clip open to form an L shape and attach the curved part of the paper clip to the cardboard with a push pin. The metal of the paper clip should touch the terminal pins, but not stop the cork from spinning. Bend and attach the other paper clip to the cardboard with a push pin so the paper clip touches the second terminal pin. These are the brushes of your motor.

7. Have an adult remove about 1 inch (2.5 cm) of insulation from each end of the two short pieces of wire. Attach the end of one wire to one of the paper clips and one end of the other wire to the other paper clip.

8. Place a horseshoe magnet over the top of the cork assembly. If the magnet is not large enough, you can place two bar magnets across from each other on the cardboard with the cork in between. They should be placed on their sides so that one magnet has its south pole facing the cork in the air and the other magnet has its north pole facing the cork.

9. Attach the free end of one short insulated wire to the positive (+) terminal of the battery and the free end of

the second short insulated wire to the negative (–) terminal of the battery. Give the cork a light spin to get it started, and watch it spin.

WHAT HAPPENED

Your motor began to turn. When you connected the wires to the battery, you formed an electric circuit. The electricity flowed from the battery, through the paper clip, into the straight pin, and into the coil. It completed the circuit by flowing out of the coil, through the pin, into the other paper clip, and back through the wire to the battery. The electric current in the coil turned the coil into an electromagnet. It was attracted to the bar magnets (or horseshoe magnet), which caused the cork to turn. When the cork turned, each pin rotated to touch the other paper clip, and the electric current changed direction. This caused the electromagnet to change polarity and become attracted to the other poles of the magnets. This keeps the coil turning in the same direction. The rotation of a motor can be used to power other machines.

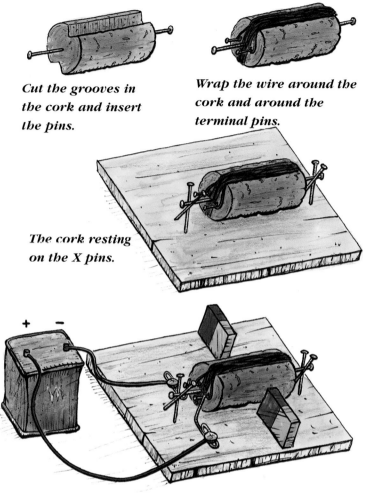

Cut the grooves in the cork and insert the pins.

Wrap the wire around the cork and around the terminal pins.

The cork resting on the X pins.

The finished motor, attached to the battery terminals.

Electric Generator

YOU WILL NEED

✔ 2 pieces of insulated copper wire, each about 25 feet (8 meters)*

✔ tall, thin empty glass or plastic bottle or large round cardboard tube*

✔ wire stripper or craft knife

✔ masking tape

✔ compass

✔ thin cardboard tube, like that used for toilet paper

✔ bar magnet

You could use the solenoid you made in Merry-Go-Rounds, if you still have it, instead.

WHAT TO DO

1. Have an adult remove the plastic insulation from both ends of one piece of insulated wire, so that about 1 inch (2.5 cm) of copper is exposed at each end.

2. Tightly wrap the wire around the bottle or large tube, so that it circles the tube at least 40 times. Leave about 1 foot (30 cm) of wire free at each end. You have made a solenoid.

A s with the electric motor, the principles that make an electric generator possible were discovered by English scientist Michael Faraday. In 1825, Faraday was made the director of the laboratory of the Royal Institution in London. While he was working there in 1831, he built the world's first **electric generator.** An electric generator converts

mechanical power into electrical power. His generator had a disk of copper, which could be turned to spin in a magnetic field. It produced a continuous electric current. This type of current is called **direct current** or DC. Here's how to make a generator that will produce a current that changes direction, called **alternating current** or AC.

3. Tape the coil of wire to the bottle or tube to keep it in place.

4. Lay the compass inside the solenoid. This will be your current detector, or **galvanometer.**

5. Wrap the second piece of wire around the thin cardboard tube to form a second solenoid. Make sure that the free ends of the wire extend outward about 2½ to 3 feet (75 to 90 cm).

6. Attach one free end of the wire from the galvanometer to one free end of the wire around the second solenoid. Attach the other free end of the wire from the galvanometer to the other free end of the solenoid wire. This makes a complete circuit. Hold the second solenoid away from the galvanometer so that it will not be affected by the magnet used in the next step.

7. Slide a bar magnet into the middle of the cardboard

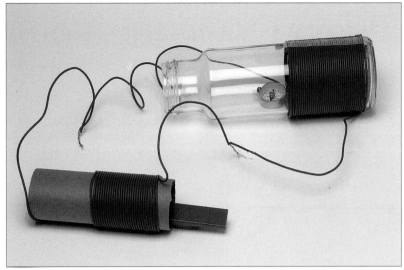

The galvanometer connected to the solenoid with the magnet inside it.

tube of the second solenoid; then pull it back out. What happens to the compass needle? Keep sliding the magnet in and out of the tube and watch what happens.

WHAT HAPPENED

Whenever the magnet moved into or out of the second solenoid, it caused an electric current to flow. This caused the needle of the compass to move. When the magnet was

not moving, no current was formed. Just as a moving electric current can cause a coil to become a magnet, a moving magnet can cause electricity in a coil. The direction of the electric current changed as you changed the direction in which the magnet traveled. An electric current that changes direction is called alternating current, or AC. Your household appliances run on alternating current.

Pump Down the Volume

YOU WILL NEED

✔ old audiocassette or video-cassette (don't use one you want to play again)
✔ audiocassette player or videocassette recorder
✔ bar magnet

If there were no magnets, you wouldn't be able to watch the latest movies on your videocassette recorder (VCR) or hear your favorite music on an audiocassette recorder. Unfortunately, you won't be able to see movies or hear the music if your magnets get too close to these pieces of equipment, either. Let's look at what happens when you let a magnet make the music.

WHAT TO DO

1. Take an old audiocassette or videocassette and play it for several seconds in the cassette player or VCR. Remove the tape from the machine without rewinding it.

2. Move it away from any other sound equipment, video equipment, or computer equipment. Pull out about 12 inches (30 cm) of tape from the cassette, but don't twist or break it.

3. Take a bar magnet and rub it back and forth over the exposed videotape or audiocassette tape, right on the surface of the piece of tape.

4. Carefully rewind the tape into the cassette.

5. Place the tape in the audiocassette player or VCR, play it, and observe what happens.

WHAT HAPPENED

The magnet changed the audiocassette or videotape. The signal on the tape is made of magnetized material arranged in a specific pattern. When a magnet is brought near the tape, the magnetic particles on the tape are mixed up. When this happens, the particles no longer have a pattern that can be read by your recorder. The recorder now just makes random noise, instead of beautiful music or images.

Sounds Like Teen Spirit

Many of you have probably tape-recorded your voice or even some music. It seems like magic when sound from one source can be reproduced exactly on a thin piece of plastic. In Pump Down the Volume, you learned how to erase a magnetic tape; in this experiment you will learn how tape recordings are made, using magnets.

YOU WILL NEED

✔ a blank audiotape cassette
✔ steel knitting needle with a sharp end
✔ about 2 yards (2 m) insulated copper wire
✔ wire stripper or craft knife
✔ 6-volt lantern battery
✔ tape recorder or audiocassette player

WHAT TO DO

1. Have an adult trim the plastic from the ends of the insulated wire, so that several inches (6 or 8 cm) of copper are exposed.

2. Tightly wrap the wire around the knitting needle, so that it coils around the needle about 40 times. Leave about 12 inches (30 cm) of wire free at each end.

3. Pull several feet (a meter) of tape out of the audiotape, making sure it doesn't get tangled. Do not cut the tape from the cassette.

4. Hold the sharp end of the knitting needle against the flattened tape. Have a helper feed the tape under the needle.

5. Connect and disconnect the ends of the wires from the terminals of the battery while your helper is feeding the tape under the needle. You can use longer and shorter connections to make a pattern.

6. Carefully rewind the tape into the cassette. Make sure

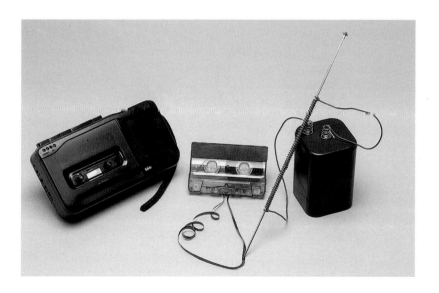

not to tangle or twist the tape as you are putting it back into the cassette.

7. When the tape is back in the cassette, put it into your tape recorder and play the section on which you experimented.

WHAT HAPPENED

You made your own tape recording using magnetism. The tape contains magnetic material. When you connected the wire to the battery, you made the knitting needle into an electromagnet. The magnetic field of the needle changed the arrangement of the magnetic material on the tape. When the tape was played, the player changed the new magnetic arrangement into sounds.

Magnetic Subs

I n Field of Streams, you saw the magnetic fields in three dimensions. In Merry-Go-Rounds, you learned how a magnetic field is created inside a coil of wire when electricity goes through it. What would happen if you combined these two experiments? This interesting activity will amaze your family and friends and would make a terrific science fair project.

YOU WILL NEED

- ✔ 10 yards (9 meters) insulated copper wire
- ✔ wire stripper or craft knife
- ✔ masking tape
- ✔ long, narrow glass or plastic jar (the kind used for sweet pickles, olives, or cherries)
- ✔ water
- ✔ small, flat piece of cork*
- ✔ small iron nail
- ✔ 6-volt lantern battery

*A slice of a wine cork will do.

WHAT TO DO

1. Have an adult trim the plastic ends from the insulated wire, so that several inches (about 6 cm) of copper are exposed at each end.

2. Tightly wrap the wire around the middle of a jar, so that it circles the jar at least 40 times. Leave about 1 foot (30 cm) of wire free at each end. Tape the wire to the jar to prevent it from sliding and unwinding.

3. Fill the jar with water so that the water level is above the coils.

4. Cut a piece of cork smaller than the opening in the jar, and push a nail through the center of the cork. Put

the cork in the bottle. The cork and nail should float on top of the water. This is your "diver."

5. Attach one end of the wrapped wire to the positive terminal of the battery and the other end to the negative terminal of the battery. Don't leave the wire connected for more than a few seconds at a time. Watch what happens to the cork and nail.

6. Connect and disconnect the wires from the battery several times, each time observing the diver.

WHAT HAPPENED

Each time electricity flowed through the coils, the diver sank towards the bottom of the jar. The coil of wire around the jar acted like a magnet each time the wires were connected to the battery. The nail was attracted by the magnetic field of the coil and moved downwards in the jar. When the electric current stopped, the magnetism of the coil disappeared and the nail could float back up to the top of the water in the jar.

The diver in the bottle when the current is not flowing.

The diver in the bottle with the current flowing.

PART THREE

Magnetic
Magic & Games

Curling Iron

YOU WILL NEED

✔ small colored plastic dessert plate
✔ plastic-coated steel or iron paper clips or other iron or steel objects such as ball bearings, washers, or bobby pins
✔ circle, bar, or horseshoe magnet
✔ strong tape

WHAT TO DO

1. Tape the magnet to the top (serving side) of the plate.

2. Turn the plate upside down. Scatter paper clips or other small metal objects over the surface of the plate that is now facing up. Do not use any sharp or pointed iron shapes.

3. Use your fingers to pull the clips or other objects into different designs.

WHAT HAPPENED

The designs you built by arranging the paper clips or other objects kept their shapes. The magnet placed on the other side of the plate was strong enough to attract the paper clips through the plastic. The objects on the plate became temporarily magnetized and attracted the objects on top of them.

*N*ow that you know the basics about magnets, here are some fun and silly activities to let you put into practice the theories you have learned. This first project, a magnetic sculpture, is based on the idea of induced magnetism we explored in Promises and Inducements.

Hidden magnet below the plate attracts paper clips.

An Extremely Charming Snake

How would you like to magically charm a snake to rise from a basket? With a few simple materials, you too can become a fakir, a person who can perform magical feats. This is a great experiment to perform at school in front of your class.

YOU WILL NEED

✔ tissue paper
✔ crayons or felt-tipped pens
✔ scissors
✔ strong glue or tape
✔ steel paper clip
✔ small circular magnet, the type found on the back of magnetic refrigerator ornaments
✔ piece of wooden doweling the same diameter as the magnet and about 12 inches (30 cm) long, or an old recorder (musical instrument).

WHAT TO DO

1. Draw the shape of a coiled snake like the one in the photo onto a large piece of tissue paper. Be sure to give the snake a nice big head. Color it with crayons or markers if you like. Cut out the snake with a pair of scissors.

2. Glue a steel paper clip to the underside of the snake's head so that it can't be seen from above.

3. Glue a small circular magnet onto the end of a piece of wooden doweling. Color the doweling so it blends in with the color of the magnet. This is your "flute." If you have an old recorder, glue or tape the magnet to the end of the recorder to create a real "flute."*

4. To charm your snake, bring the end of the flute with the magnet attached near the head of the snake and slowly move the flute upwards, allowing the snake to follow. If you are using a recorder, you can play a tune while the snake rises.

*Get your parents' permission before you do this to your recorder.

The recorder and paper snake.

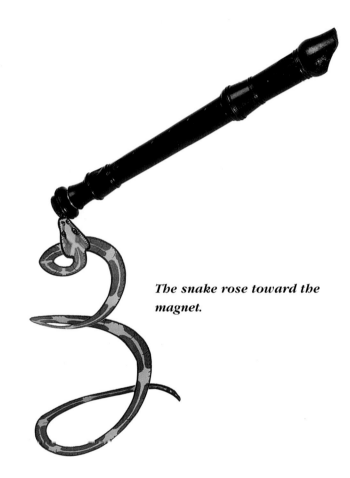

The snake rose toward the magnet.

WHAT HAPPENED

The magnet attracted the paper clip underneath the snake's head. The attraction was strong enough to lift up the lightweight paper snake, making it appear to move off the table on its own.

Gone Fishin'

WHAT TO DO

1. Draw and cut out 12 to 24 fish from paper, cellophane, or cardboard. These fish can be just about any size, but must fit in the container.

2. Write any number from 1 to 10 on the back of each fish.

3. Attach a paper clip to the middle of each fish, and place the fish in the container.

4. For each fishing rod, tie a piece of string about 8 inches (20 cm) long to one end of a chopstick; tie a magnet to the opposite end of the string.

5. Use the magnetic fishing rods to catch fish. If you are playing this game with a friend, record the number on the back of each fish, and see who has the most points at the end of a fishing round.

WHAT HAPPENED

The magnet on the end of the rod attracted the paper clip, picking up the fish.

*H*ere's a simple way to go fishing without a boat, a fishing license, or even a reel. Best of all, you won't have to throw the small ones back!

YOU WILL NEED

- ✔ paper, colored cellophane, or light cardboard
- ✔ crayons
- ✔ scissors
- ✔ steel paper clips
- ✔ several chopsticks or other thin sticks
- ✔ string
- ✔ several small magnets (one for each fisher)
- ✔ flat nonmetal container, such as a paper plate

Animal Magnetism

Have you ever wondered how monarch butterflies migrate each year to the same place or how salmon return to spawn at the place where they were born? Scientists think that magnetism plays a part in these phenomena. It is thought that some creatures have "animal magnetism" and are affected in different ways by magnetic fields. Scientists studied homing pigeons by placing small electromagnets on their heads. These pigeons tended to fly in the wrong direction or opposite to the way in which they normally flew, when the polarity of the magnet was reversed. Tiny crystals of magnetite have also been found in the human brain, but you can't use them as a compass to find your way home. Animals aren't the only living things with magnetite: some bacteria containing small grains of magnetite will align themselves with a magnetic field.

Keeping Afloat

*H*ere are two neat tricks you can do to amaze your friends with your magnetic personality.

TRICK 1:
Flying Paper Clip

Bet a friend you can lift a paper clip without letting anything touch it. Not only that, but it will hang in midair. Here's how.

YOU WILL NEED

✔ magnet
✔ string, about 6 inches (15 cm)
✔ steel paper clip
✔ book or other heavy object

WHAT TO DO

1. Tie one end of the string to the paper clip and place the other end of the string under a heavy book. Dangle the string and paper clip off the side of a table.

2. Hold the magnet just far enough away from the end of the paper clip so that it does not touch the clip, but will still attract it.

3. Lift the paper clip up using magnetic attraction.

WHAT HAPPENED

The paper clip was attracted to the magnet and rose off the side of the table.

TRICK 2:
Floating Magnets

Now bet your friend that you can suspend a magnet in air without touching it. Here's how.

YOU WILL NEED

- ✔ 3 identical circular magnets with holes through their centers
- ✔ a wooden or bamboo stick
- ✔ modeling clay

WHAT TO DO

1. Place a blob of modeling clay on a flat surface such as a tabletop.

2. Poke one end of the stick into the clay so that it points upwards.

3. Slide one magnet, north pole facing downward, onto the stick.

4. Drop the second magnet, south pole facing downward, on top of the first one.

5. Drop the third magnet, north pole downward, on top of the other two magnets.

WHAT HAPPENED

The second and third magnets floated or seemed to levitate one above the other on the stick. The like poles repelled each other, pushing the magnets apart.

"Floating" magnets.

Clip Art

YOU WILL NEED

✔ small cookie sheet, rectangular metal cake pan, or other rimmed container
✔ paper
✔ tempera paints
✔ magnet
✔ small metal objects that are attracted to a magnet, such as steel paper clips, ball bearings, washers, nuts and bolts, and nails

I s it possible to finger-paint without using your fingers? Here's a way to create some really unique art, using magnetic attraction.

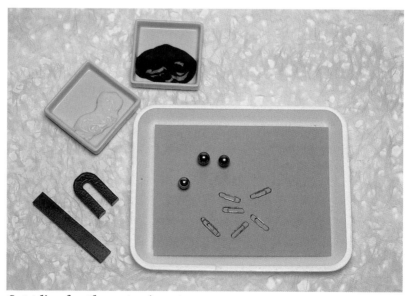

Supplies for the experiment.

WHAT TO DO

1. Place the paper on the bottom of the container and drizzle several spoonfuls of different-colored paint onto the paper.

2. Drop one or two of the small objects onto the paint.

3. Place the magnet underneath the container and use it to attract the objects. Slide the objects through the paint to create interesting designs. You can use several metallic objects to paint your picture.

4. Carefully remove the objects from the container and rinse them with water to remove the paint. Remove your paper and lay it on a flat surface to dry.

A magnetic painting.

WHAT HAPPENED

You have painted a magnetic masterpiece suitable for hanging on any refrigerator!

Receding Hairline

This is a silly science experiment, based on a popular toy. One child we know cut out a photograph of her father, who was slightly balding, and used it as the picture for her magnetic art. You might find it amusing to give your mother a mustache, or your baby brother or sister could have a beard.

YOU WILL NEED

✔ picture from a magazine, a photograph,* or a hand-drawn picture
✔ glue
✔ paper plates (1 for each picture)
✔ plastic wrap
✔ cellophane tape
✔ iron filings
✔ magnet

*Be sure you get permission before cutting up any photos.

WHAT TO DO

1. Glue the picture to the paper plate. It should fit inside the bottom of the plate.

2. Cover the picture with clear plastic wrap. Smooth the wrap down; then tape the wrap to the rim or back of the plate.

3. Place several teaspoons of iron filings on the plate.

4. Place the magnet below the plate and move the iron filings around to create "hair," "mustaches," or other funny features.

5. If you wish, you can create your own picture by drawing on the plate.

6. When you are finished, carefully remove the filings from the plate and return them to their container.

WHAT HAPPENED

The magnet attracted the iron filings through the paper plate. As you moved the magnet around, the filings became the features on your picture. When the magnet changed position, so did the filings.

Photos of the authors.

Glossary

alternating current: an electric current that regularly reverses its direction.

battery: two or more electrochemical cells connected together. It converts chemical energy into electrical energy.

compass: a device for finding directions by means of a small magnet turning on a pivot, which points to the Earth's magnetic north.

Curie point: the temperature above which a ferromagnetic substance will lose its magnetism and become less magnetized (paramagnetic). For iron the Curie point is 770°C. Discovered by French scientist Pierre Curie in the 1890s.

degree (abbreviated °): (1) A unit of measure based on divisions of a circle. A full circle is 360°. A quarter circle is 90°. (2) One of the divisions or intervals on a scale of a measuring instrument, such as a thermometer.

diamagnetism: a kind of magnetism that is in the opposite direction to the magnetic field. Materials that have this quality are called diamagnetic.

dip needle: a magnetic needle used to measure the vertical part of the Earth's magnetic field, an angle with the horizontal plane called the angle of inclination.

direct current: an electric current in which the flow of the charge is in one direction only.

eddy current: an electric current induced in a conductor that is moving in a magnetic field.

electric circuit: the complete path through which an electric current flows.

electric current: a flow of electrons through a conductor, such as a wire, or of charged particles through an electrolyte.

electric generator: any machine that converts mechanical power into electric power. Electromagnetic generators may be driven by steam turbines, water turbines, windmills, etc.

electromagnet: a coil of wire with an iron core, which acts as a magnet when an electric current flows through the wire and loses its magnetism when the current ceases.

electron: an atomic particle with a negative charge. Flowing electrons make up an electric current.

ferromagnetic material: a substance like iron or steel that is strongly attracted to a magnet and can keep its magnetic charge.

galvanometer: an instrument for detecting and measuring small electric currents.

geomagnetism: the natural magnetism of the planet Earth, which acts like a giant magnet.

inclination (magnetic dip): the angle that a freely suspended magnet will make with the horizontal, as it points downwards due to the Earth's magnetic attraction.

induced magnetism: temporary alignment of the magnetic domains in a magnetic material that is placed near a magnet.

lines of force: lines in a field of force around a magnet, which can be drawn around the magnet to show the direction of the magnetic effect.

lodestone: a naturally magnetic rock made of the mineral magnetite.

magnetic attraction: a force between a magnet and an object, pulling the object towards the magnet.

magnetic domains (or *domains):* very small (1 to 0.1 mm) magnetized regions of a ferromagnetic substance. In a strong magnetic field, all the domains are lined up in the same direction as the field.

magnetic field: a field of force that exists around a magnet or a current-carrying conductor.

magnetic poles: the areas at the ends of a magnet where the magnetic forces are strongest.

magnetic repulsion: the force with which a magnet and another object (or particles) push each other apart.

magnetite: a mineral that is naturally magnetic. Lodestones are made of magnetite, which is mostly made up of an iron oxide.

motor: any device for converting chemical or electrical energy into mechanical energy.

north-seeking pole (called *north pole* for short): the pole of a magnet that is attracted to Earth's north magnetic pole.

residual magnetism: magnetism retained by a magnetized object after a magnet is no longer near the object.

right-hand rule: a rule used to describe the direction of the magnetic lines of force around a wire with an electric current.

solenoid: a coil of wire wound on a cylinder in which the length of the cylinder is greater than its diameter. When a current is passed through the coil, a magnetic field is produced inside the coil parallel to its axis.

south-seeking pole (called *south pole* for short): the pole of a magnet that is attracted to the Earth's south magnetic pole.

Index

alternating current, 60, 61
Aluminum Slidings, 56–57
Ampère, André Marie, 51
angle of inclination, 35–36
animal magnetism, 71
artificial magnets, 31
atoms, 29
attraction, 13
audiocassette, 62, 63

bar magnet, 17, 23–24, 25, 26
battery, 6, 7, 45, 47

caring for and storing magnets, 38
Chain Gang, 26–27
circular magnets, 21
Clip Art, 74–75
compass:
  affected by magnet, 24
  deflected by electric current, 43–44, 50
  demagnetizing, 7
  early, 10, 32
  how to use, 11–12
  making your own, 33
computer equipment, and magnets, 7
Curie point, 40
Curling Iron, 67
current:
  effect on compass, 43–44, 50
  in electromagnet, 47, 53–54
  flow of, 47
  right-hand rule, 44

demagnetizing, 7, 19, 28, 38, 39–40
diamagnetic substance, 29, 41
dip needle, 35–36
"diver," 65
domains, 28, 29, 38
dry cells and wet cells, 45
Due South, 32–33

Earth:
  as magnet, 12, 15–16, 33, 34–35, 37, 38
  magnetic field of, 35, 36, 56
  magnetic model of, 34–35
eddy currents, 56–57
Electric Generator, 60–61
Electric Motor, 57–59
electrolyte, 45
electromagnet, 31, 50, 51, 52, 53–54, 55
electrons, 29, 47
equipment and supplies, 8
Extremely Charming Snake, 68–69
Eye of the Needle, 31–32

Faraday, Michael, 57, 60
Faster, Higher, Stronger, 51–52
ferromagnetic substance, 29, 39
Field of Streams, 22–23
fishing game, 70
Floating Magnets, 73
flow of a current, 47
Flying Paper Clip, 72

galvanometer, 61
generator, 60–61
Get a Lode of This, 10
Gilbert, William, 34
glossary, 77–78
Going Round in Circles, 48–49
Gone Fishin', 70

Hammerheads, 37
heating a magnet, 39–40
horseshoe electromagnet, 55
horseshoe magnet, 18, 20, 27, 38
Hot Shoes, 54
Hot Stuff, 39
Hwang-ti, Emperor, 10

I'm So Inclined, 35–37
inclination, 35–36
induced magnetism, 30
iron filings:
  magnetizing and demagnetizing, 28
  pictures made with, 76
  to show magnetic field, 17, 18, 20, 22, 46
iron rod, 31

Keeping Afloat, 71–73

light-sensitive paper, 8, 20–21
lines of force, properties of, 25
lodestone, 10, 32

Magnetic Attractions, 13–14
magnetic domains, 28, 29, 38
magnetic field:
  around a wire, 44, 45–46,
    48–49, 50, 51, 54, 65
  of bar magnet, 16–18,
    23–24, 25, 26
  of Earth, 35, 36, 37, 56
  of horseshoe magnet, 18,
    20, 27, 38
  right-hand rule, 44
  shielding an object from, 19
  three-dimensional, 22
magnetic flute, 68–69
magnetic north, 12
magnetic painting, 74–75
magnetic poles, 16, 17, 27
magnetic repulsion, 13
magnetic sculpture, 67
Magnetic Subs, 64–65
magnetite, 32
Maricourt, Pierre de, 16
Merry-Go-Rounds, 49–50
metals attracted by magnets,
  14, 17, 19
motor, 57–59

nail, as temporary magnet, 30

needle, how to magnetize,
  31–32
nonmagnetic metals, 14

Oersted, Hans Christian, 43
On the Level, 40–41

paramagnetic substance, 29
Playing Fields, 16–18
Promises and Inducements,
  30–31
Pump Down the Volume, 62

Receding Hairline, 76
repulsion, 13, 18, 73
residual magnetism, 30
right-hand rule, for current in
  wire, 44
Round and Round, 15–16

science fair projects, 6
Shake It Up, 27–28
sharp tools, 6
shielding an object from
  magnetic field, 19
Shields Up, 18–19
Slim Pickin's, 46–47
solenoid, 50, 61, 65

Sounds Like Teen Spirit,
  63–64
storing magnets, 38
striking metal, to realign
  domains, 38

Tale of Hans Christian
  Oersted, 43–44
Tan Lines, 20–21
temporary magnet, 30
terminal, of battery, 45
tools and supplies, 8
Two Are Better Than
  One . . . Or Are They? 53

Using a Compass, 11–12

Vikings, 32

watches, and magnets, 7
What a Wonderful World,
  34–35
Which Way Did It Go? 23–25
wire, magnetic field around,
  44, 45–46, 48–49, 50, 51, 54,
  65
wire holder for magnet, 15